GOLD FEVER

America's First Gold Rush

by RAY C. RENSI *and* H. DAVID WILLIAMS

GEORGIA HUMANITIES COUNCIL

Georgia History and Culture Series

Georgia Humanities Council
Atlanta, Georgia 30322

Printed in the United States of America

The Georgia Humanities Council receives support through grants from the National Endowment for the Humanities. The views herein do not necessarily represent those of the Georgia Humanities Council and the National Endowment for the Humanities.

Contents

About This Publication

Gold fever came to Georgia over two decades before the famous California gold rush. With the excitement and sudden riches for some came environmental choices, competition for control of land, and harsh treatment of the Native Americans who inhabited the region. The story of Georgia's gold rush is fascinating in itself; it also offers an opportunity to consider some issues that are very much with us today.

This publication is intended as a study guide for viewers of "Gold Fever," a half-hour video documentary sponsored by the Georgia Department of Natural Resources and produced and directed by Jim Couch and Becky Marshall. The film and the guide taken together form a unit in a series, "Georgia History and Culture," developed by the Georgia Humanities Council with financial support from the National Endowment for the Humanities. The series is designed to make available to the general public the results of scholarly research on Georgia and the South in the fields of study known as the humanities. Among these fields are history, archaeology, literature and language, philosophy, and ethics, all of which are basic in school and college education. The contributions of specialists in the humanities and their writings are indispensable resources for interpreting our cultural heritage and reexamining our values.

The video documentary tells the story of the gold rush in Georgia, and includes the recollections of some local citizens. The publication offers a more detailed historical account and analysis of the gold rush and its effects and also includes suggestions for further reading and questions for discussion. They may be used by student and adult groups, alone or as part of a series of discussions. A discussion leader may draw on this publication as a resource in preparing a program and may also make it available to participants.

The discussion questions provided are open-ended, and are intended to suggest ways of fostering exploration of the issues. Members of your group will undoubtedly bring their own knowledge, experience, and interests to the conversation. You should not feel limited by the questions provided. Among the broad issues arising from this story are the environmental impact of mining, the rights of the Cherokees, and the powerful attraction of gold.

In the suggestions for further reading, we have tried to include items that will be readily available to the general reader in local libraries. For those who may wish to explore further, we have also included some references which may require a visit to a larger public or academic library. You will want to search your local library, as well, for other books and magazines that deal with the general issues and questions mentioned above.

We hope that these materials will stimulate the development of programs and series of programs in local libraries, historical societies, church and civic groups, and other organizations. Arrangements to borrow a videotape copy of the film or to obtain copies of this guide may be made by contacting the Georgia Humanities Council, 1556 Clifton Road N.E., Emory University, Atlanta, Georgia 30322; telephone (404) 727-7500. Information is also available about other programs in this series, and about grant programs of the Georgia Humanities Council through which groups may seek support for educational programs.

Mary Drake McFeely
Series Editor

The Lure of Gold

Said Benjamin Parks of gold mining, "It is just like gambling—all luck."

On a crisp autumn day in the mountains of northern Georgia, young Benjamin Parks was returning home from a "lick log" on the western side of the Chestatee River where he and his friend, Lew Ralston, provided salt for their cattle. The deer were in abundance this time of year and "Uncle Benny," as he was know to the locals in later years, strolled quietly through the woods hoping to catch sight of a possible supply of meat on the hoof. While engaged in thoughts of deer steak, he accidentally kicked up a stone. The color of this otherwise common stone caught his attention and he paused for a closer look. As he knelt and held it in his hand, its true nature became clear to him. Years later, he would remember that the color was something like that of an egg yolk—a deep rich yellow hue. It was gold.

Benjamin Parks was not the first to be captivated by the luster of gold in the southern Appalachians. The first people to inhabit the region were descendants of nomadic bands of hunter-gatherers who migrated to the western hemisphere, perhaps twenty thousand years ago, across the land bridge that existed between what are today Alaska and Siberia during the last ice age. There is little evidence that these original Americans sought gold in significant quantities. The yellow metal does not seem to have held the same power over these people that it had over subsequent generations of Europeans, but the Indians did find gold attractive as jewelry. These adornments were often used as items of trade in an extensive network that included all of the present-day United States and may have extended as far south as the Yucatan Peninsula in what is today southern Mexico.

The first Europeans to touch the shores of North America in search of gold were the Spaniards. In 1513, Ponce de Leon was informed by the Florida natives of a land to the north where gold might be found in abundance. The first real evidence of gold came to the Spanish a few years later when they were presented with a small sampling said to have come from a province deep in the interior called Apalacha. The promise of gold lured the Spanish northward and in 1540 an expedition was organized under the conquistador Hernando de Soto. In his travels through the gold region of the southeast, he heard fantastic tales of natives who wore golden hats and of lands where gifts of gold were given in great quantities to women who ruled them. Some accounts place de Soto in Nacoochee Valley where he is said to have been received by the Indians with hospitality. But de Soto's quest was in vain, as were those of most other Spaniards who followed him in later years. It is difficult to know for sure how much gold the Spanish were able to glean from their workings in what would become northern Georgia, but it seems evident from the excavations of Spanish mining operations and the tenacity with which they resisted British and French encroachment that the Spaniards held to the notion of a North American El Dorado for as long as their weakening international position would allow.

The British, though later arrivals, were no less eager in their quest for gold. The first permanent English settlement at Jamestown nearly failed because of a fruitless gold fever that struck the village in 1608, a year after its founding. According to Captain John Smith, "There was no talke, no hope, nor worke, but dig gold, wash gold, refine gold, load gold." But the newcomers were no more successful than the Spanish had been.

Over the next two centuries there was periodic excitement over the prospect of big strikes, but none really "panned out." A few mines were worked in North and South Carolina prior to the American Revolution, and Thomas Jefferson, in his *Notes on Virginia,* told of a piece of gold ore weighing about four pounds that was found on the Rappahannock.

The first discovery to result in the establishment of permanent mining operations in the United States occurred in 1799 in Cabarrus County, North Carolina, on the property of a farmer named John Reed. Reed's son, Conrad, came across a speckled stone glittering with small nuggets while archery fishing along Little Meadow Creek. It was a seventeen-pound lump, but the Reeds saw it only as a curiosity and used it as a doorstop. Three years later, Reed took the rock with him on a trip to Fayetteville where, to his astonishment, it was identified as gold ore. Word of the find spread slowly, but by 1804 other farmers in the area had found gold on their own lands and regular mining works were set up.

The North Georgia Gold Rush

No one is certain when the first discovery of gold was made in northern Georgia. Perhaps the Cherokee knew there was gold on their lands, but they were understandably reluctant to have that fact generally known. Even so, the secret could not be kept indefinitely and by the late 1820s white men were beginning to realize that the natural beauty of these Appalachian foothills concealed hitherto unknown wealth.

Long before the whites knew of gold in the region, they coveted the lands of the Cherokee Nation. In 1802, Georgia gave up its claims to the areas that are today Alabama and Mississippi in exchange for a promise from the federal government to remove all Indians from her remaining claims "as soon as the same can be peaceably obtained on reasonable terms." Georgia's efforts at Indian removal were directed first against the Creeks who inhabited the rich cotton land on the east bank on the Chattahoochee. By 1825 the Treaty of Indian Springs sealed the fate of the Creeks and Georgia quickly turned its attention northward to the Cherokee Nation, bounded on the south by the Chattahoochee and on the east by the Chestatee.

The Cherokee had long since determined to preserve their homeland by adopting many aspects of white culture. They became settled farmers, built towns, engaged in trade and commerce, developed a syllabry for their language, adopted Christianity, and formed a Cherokee government based on the Constitution of the United States. By 1826, the Cherokee had come to be referred to as one of the five "civilized" tribes.

Nevertheless, in December of that year the Georgia State Senate passed a resolution requesting that the president of the United States take steps to initiate a treaty with the Cherokee, "the object of which shall be to extinguish the title to all or any part of the lands now in their possession within the limits of Georgia." The federal government was reluctant to act on Georgia's request, since it was bound by previous treaty obligation to protect the Cherokee from further white encroachment. The 1802 treaty with Georgia stipulated that any lands granted by the Cherokee must be "peaceably obtained," and the Cherokee obviously had no intention of giving up their nation without a struggle.

Impatient with federal authorities and alarmed at the entrenchment of the Cherokee, Georgia took matters into her own hands. On December 20,

1828, the state legislature enacted a bill mandating the extension of Georgia's authority over the Cherokee, to take effect on June 1; 1830. After that date all laws and customs of the Cherokee Nation would be declared null and void.

The year 1828 was also a historical landmark for Georgia in another respect. Several nearly simultaneous discoveries of gold were made at sites on or near Cherokee lands, and by the end of the year the first genuine gold rush in United States history was underway. A slave owned by a Major Logan found gold in a branch of the Nacoochee River on the Lovelady place. Logan and his servant were returning from the gold fields around Rutherford, North Carolina when the slave noticed a similarity between the soil of the Nacoochee valley and that of the gold region to the north. He tested a sample and discovered that it indeed contained gold. At about the same time another black man found gold on Bear Creek near the present site of Dahlonega. Jesse Hogan, a prospector from North Carolina, claimed to have found gold on a branch of Ward's Creek, also near Dahlonega.

"The color was something like that of an egg yolk—a deep rich yellow hue. It was gold."

Thomas Bowen made yet another discovery on Duke's Creek in what is today White County. According to Judge John Underwood, to whom Bowen related his story, "There came up a storm and blew down some tall timber along Duke's Creek . . . and in the roots of the timber Mr. Bowen found . . . gold."[1] Another Duke's Creek find was made by John Witherow, or Witheroods, who came across a three-ounce nugget in the little mountain stream. And Benjamin Parks made his find just east of the Chestatee in what was then Hall County but would four years later become part of Lumpkin.

The only discovery for which a date is claimed is that of Benjamin Parks. Said Parks, "It was my birthday [October 27], so I'd ought to know. And I'm a-telling you, the to-do that followed was going full steam before ever 1829 was rung in."[2] But despite the popularity of the Parks legend, no firm eveidenc appears until 1 August 1829, when the *Georgia Journal* reported that two gold mines had been discovered in Habersham County.

In any case, by late 1829 the Cherokee Nation was being invaded by prospectors hungry for gold. It is difficult to know exactly how many people were involved in the gold rush, but clearly they numbered in the thousands. *Niles' Register* reported that in May of 1830 there were four thousand miners washing the sands on Yahoola Creek alone. As an old man in his nineties, Uncle Benny recalled the scene.

> The news got abroad, and such excitement you never saw. It seemed within a few days as if the whole world must have heard of it, for men came from every state I had ever heard of. They came afoot, on horseback and in wagons, acting more like crazy men that anything else. All the way from where Dahlonega now stands, to Nuckollsville there were men panning out of the branches and making holes in the hillsides.[3]

George Gilmer, then Governor of Georgia, also related his impression of these "Twenty-Niners" in the gold region.

> Many thousands of idle, profligate people flocked into it from every point of the compass, whose pent up vicious propensities, when loosed from the restraints of law and public opinion, made them like the evil one in his worst mood. After wading all day in the Etowah and Chattahoochee Rivers, picking up particles of gold, they collected around lightwood knot fires, at night, and played on the ground and their hats, at cards, dice, push pin, and other games of chance, for their day's findings. Numerous whiskey carts supplied the appropriate aliment for their employments. Hundreds of combatants were sometimes seen at fisticuffs, swearing, striking and gouging, as frontier men only can do these things.[4]

The spread of gold fever was fueled by daily reports from the mines of spectacular success. Newspapers throughout the country carried articles telling of nuggets worth anywhere from one to two hundred dollars. Claims were made that a single miner with no more than a pan and shovel could gather upwards of thirty dollars' worth of gold per day. Actually, these early prospectors were fortunate to pan out five dollars a day (the price of gold at the time was close to twenty dollars an ounce), which wasn't bad considering that the average salary for a worker employed by mining companies a few years later was around $1.50 per day. Even so, a significant amount of gold was coming out of north Georgia. In 1830, the United States mint at Philadelphia received $212,000 from the Georgia gold fields and by the end of 1832 over half a million dollars worth had been deposited.

"All the way from where Dahlonega now stands to Nuckollsville, there were men panning out of branches and making holes in the hillsides."

The sudden influx of thousands of miners into the Cherokee Nation was known even at the time as the "Great Intrusion." Georgia was not to take possession of the area until June of 1830 and even then the Cherokee would retain title to their property until the state might decide how best to dispose of it. The natives protested loudly against this obvious infringement of the agreement. Wrote the *Cherokee Phoenix*, "Our neighbors who regard no law and pay no respects to the laws of humanity are now reaping a plentiful harvest . . . We are an abused people." But there was little the Cherokee could do. It seemed the more vehemently they protested, the more eagerly the miners came.

Desperate to stem the flow of intruders into their nation, the Cherokee turned to the federal government for help. The United States agent to the Cherokee issued an order for all intruders to cease their mining activities immediately and remove themselves from Cherokee territory. To enforce this order the United States dispatched army regulars in an attempt to drive out the gold diggers.

Immediately upon their arrival, the troops began making arrests. Detachments roamed the hillsides and creek bottoms in search of miners, taking them into custody, destroying their equipment, and hastily escorting them off Cherokee lands, but no sooner was one group of miners driven off than another took its place.

Sometimes the abandoned workings were taken over by Cherokee gold hunters. This infuriated the former occupants and they began returning to the mines in droves. To avoid conflict, the army adopted a policy forbidding anyone, native or intruder, to work the mines. This did little to turn back the flood of miners, and the arrests continued. All the way from the mines along "Sixes" Creek in what is now Cherokee County, northeast to the Chestatee River, men were rounded up and driven off, only to return when their evictors moved on to the next mining camp. Major Phillip Wager, commanding the federal force, recalled the results of a single night's arrests at the "Chestatee Mines."

> The collection made . . . of upwards of 200 persons presented a most motley appearance of whites, Indians, halfbreeds and negroes, boys of fourteen and old men of seventy—and indeed their occupations appeared to be as various as their complexions comprising diggers, sawyers, shopkeepers, peddlers, thieves and gamblers, etc. Besides them were also found in the hopeful assemblage two Colonels of Georgia Militia, two candidates for the legislature and two ministers of the Gospel, all no doubt attracted thither by the love of gold.[5]

Governor Gilmer and the Georgia legislature howled in protest at the very notion of the United States Army exercising authority within what they considered to be the boundaries of their sovereign state. The Georgians also feared that federal protection might further steel Cherokee hearts against removal. On June 1, 1830, Gilmer issued a proclamation "notifying all persons whom it may concern, that the jurisdiction of this State is now extended over all the territory in the occupancy of the Cherokees." He then wrote a letter to President Andrew Jackson informing him of the proclamation and complaining bitterly that the continued presence of federal troops was a violation of the United States Constitution. Jackson was himself an advocate of Indian removal and had only a month before rammed through Congress the hotly debated Indian Removal Act calling for the relocation of all eastern tribes to areas beyond the Mississippi River. He complied with Gilmer's request and agreed to withdraw the army.

Gilmer immediately issued a second proclamation ordering both intruders and Cherokee to cease their mining operations on land that the state considered its own. He also appointed a Superintendent of Public Lands "whose business it was to restrain trespassers by writs of the courts." But court orders could hardly be expected to succeed where the army had failed. The prospectors simply ignored these writs and continued their gold hunting. Gilmer recognized the futility of his attempts to close the mines and later admitted, "I found paper bullets but light artillery against masses of men who could not read." Lacking authority from the legislature to send Georgia troops into the region, Gilmer reluctantly acquiesced to the presence of the United States Army until he could be empowered to raise a force of men to occupy the mines.

On December 22, 1830, the legislature passed an act creating a force of sixty men called the Georgia Guard, for the express purpose of driving gold diggers from the mines. Anyone caught prospecting by the Guard would be subject to a penalty of four years at hard labor. Haphazardly enforced, the law did little to deter men skilled in avoiding the army. The miners posted lookouts to warn of the Guard's approach and then scattered into the surrounding hills when the alarm was sounded.

Georgia and the Supreme Court

David Crockett of Tennessee voted against the Indian Removal Act of 1830. Said Crockett, "I would sooner be honestly and politically damned than hypocritically immortalized."

Besides forbidding gold hunting on state lands, Georgia also made it illegal for any white person to reside in the Cherokee country without first taking an oath to uphold the laws of the state. This act was aimed primarily at missionaries living among the Cherokee. These people were fiercely opposed to Georgia's actions and nearly as determined in their resistance as the Cherokee themselves. Only a week after the creation of the Georgia Guard, missionaries representing the Moravians, Congregationalists, Baptists, and Presbyterians met at New Echota, the Cherokee capitol, and signed a resolution supporting the Cherokee cause and declaring their decision to stay. Methodist missionaries had already done so.

The inevitable showdown between Georgia and the missionaries came in March of 1831, when the Guard arrested eleven churchmen who refused to pledge an oath to the state. At their trial, nine of the eleven agreed to take the oath, but two, Samuel A. Worcester and Dr. Elizur Butler, refused and were sentenced to four years at hard labor. Even Georgians who supported the state's acquisition of Cherokee lands were appalled by this heavy-handed treatment of the missionaries. Their only recourse was an appeal to the United States Supreme Court.

In the case of *Worcester v. Georgia*, Chief Justice John Marshall, speaking for the majority, handed down the court's decision. "The Cherokee Nation . . . is a distinct community, occupying its own territory . . . in which the laws of Georgia can have no right to enter but with the assent of the Cherokees. The Act of the state of Georgia . . . is consequently void." Worcester and Butler had won. John Ross, Chief of the Cherokee, and the rest of the Cherokee Nation received the news with overwhelming excitement. Their ancestral homeland was theirs once again.

The Cherokee's joy was short-lived, for it quickly became clear that Wilson Lumpkin, the new governor of Georgia, had absolutely no intention of abiding by the Court's ruling. The missionaries remained in the state penitentiary and the Georgia Guard remained in place to enforce the governor's will. Infuriated by this blatant disregard for the judicial process, the Cherokee sent a delegation to Washington in a last-ditch effort to obtain President Jackson's support. John Ridge, leader of the delegation, asked Jackson directly whether he intended to carry out his Constitutional obligation to enforce the Supreme Court's decision. Jackson replied pointedly that he had no such intention. Instead, he advised Ridge to return home and tell his people that their only hope for survival as a nation was to abandon their homes and move beyond the Mississippi. Upon hearing of Jackson's remarks, an elderly man named Speckled Snake, who was then more than one hundred years old, spoke to his fellow Indians.

> Brothers! I have listened to many talks from our great white father. When he first came over the wide waters, he was but a little man . . . very little. His legs were cramped by sitting long in his big boat, and he begged for a little land to light his fire on. But when the white man had warmed himself before the Indians' fire and filled himself with their hominy, he became very large. With a step he bestrode the mountains, and his feet covered the plains and the valleys. His hand grasped the Eastern and the Western sea, and his head rested on the moon. Then he became our Great Father. He loved his red children, and he said, "Get a little further, lest I tread on thee."
>
> Brothers! I have listened to a great many talks from our great father. But they always began and ended in this—"Get a little further; you are too near me."[6]

The Inheritance is Gambled Away

With a promise of noninterference from Jackson, Georgia was left with a free hand in what was now solidly its northern territory. Surveyors entered the area and began partitioning the new lands. Survey teams consisting of a surveyor, a chain carrier, a pack carrier, two axe men, and a cook, roamed the hills and valleys of north Georgia for months preparing the region to be raffled off in a statewide lottery scheduled for the autumn of 1832. They divided the Cherokee country into four sections, each containing a designated number of districts. It was also the surveyor's task to designate those districts in which gold could be found. The Gold Districts, of which there were thirty-three, were divided into lots of forty acres each, while the Land Districts, numbering sixty, were divided into lots of 160 acres.

Though most Georgians were intrigued by the prospect of a lottery in which they might literally win a gold mine, there was nonetheless considerable opposition to the idea. Many believed it a terrible waste to distribute the state's wealth in such a haphazard fashion. Gilmer himself argued that it would be more beneficial for the state to operate the mines. Taxes could be lowered, improvements on roads and navigable rivers could be made, and revenues could be generated for public education. According to Gilmer, "Mines . . . should be managed for the general and not the individual advantage."[7] This question was hotly debated in the gubernatorial campaign of 1831. Wilson Lumpkin enthusiastically supported the lottery, pointing to the need for "a settled freehold population on every part of our territory . . . hitherto the abode of a people wholly unqualified to enjoy the blessings of wise self-government."[8] Newspapers across the state took up the argument. Typical of the editorial comments was this emotional appeal from the *Macon Advertiser* of 23 September 1831:

The Question At Issue

> The two propositions which now present themselves for the decision of Georgia are: Gilmer and the Laws and Lumpkin and Taxation!!! Will the people hesitate which of the two to choose? They cannot, they will not. Let their cry be on the first Monday in October, Long Life to Gilmer! Long Life to the Laws! And Down, Down with Lumpkin and His Abominable System of Taxation!!!

But the lure of gold was too great for Georgians to overcome. They literally voted themselves a gold mine and Lumpkin became the new governor.

The state legislature designated those person who were qualified to participate in the drawing and determined the number of chances to which they were entitled. For the 160-acre land lots, qualified individuals were: "Every male white person of eighteen years of age . . . and an inhabitant . . . of this state, four years . . . Every deaf and dumb or blind person of like residence. . . All widows . . . All families of orphans." Heads of families, "families of orphans consisting of more than two," veterans of the wars with Great Britain or the Indians, and widows or orphans of such veterans were all entitled to an extra chance in the lottery. Qualifications for the 40-acre gold lots were generally the same, except that one need only have been a resident of the state for three years instead of four. Individuals who were convicted felons or had hunted for gold illegally since June 1, 1830 were excluded from participation in the lottery. The Cherokee were also excluded from the drawing by Section 16 of the lottery act, which forbade participation by any resident of the Cherokee Nation.

By September 1832, the surveys were complete. All across the state, people flocked to their respective county seats to register for the drawing. The names were sent to the state capitol at Milledgeville where they were deposited in a large wooden drum from which they would be drawn. In another drum were placed tickets identifying individual land or gold lots.

On October 22, Georgia, as one lottery opponent put it, began gambling away its rich inheritance. The twin drums were spun and names and tickets began to be drawn. In the land lottery, there were 85,000 people competing for 18,309 prizes. Blank tickets were placed in the prize drum to bring their number up to that of the number of registrants. This meant that a little more than three out of four hopeful participants won nothing more than a worthless slip of paper. In the gold lottery there were 133,000 names registered to draw for 35,000 lots. Again, as in the land lottery, about three out of four people came away empty-handed.

By April 1833 the lottery was completed and the "fortunate drawers," as they were called, began to seek out their newly acquired lands. However, these people were forbidden by the state to occupy their lots if any Cherokee lived on it. They were to wait until the Cherokee either voluntarily moved or were forced by the state to do so. For a lottery winner to attempt to coerce the original owner off his land meant forfeiture of the prize. But in fact, no sooner had many of the winners been informed of their good fortune than they rushed to take possession. Some began arriving only a few weeks after the lottery had begun. A November 24th editorial in the *Cherokee Phoenix* complained that "The fortunate drawers (so called) *of our lands* have been passing and repassing single and in companies . . . in search of the splendid lots which the rolling wheel has pictured to their imaginations. 'Ho, sir! Where is the nearest line to this place, what district, number, corner, lot, station, etc.' are the impertinent questions forced upon us." But such protests were no more effective than the Cherokees' previous efforts had been.

"Civilization" Comes to the Gold Region

Long before the drawings ended, white settlements began to spring up in the Cherokee country, particularly in the nearly uninhabited part of the gold region in what is today Lumpkin County. The first of the gold rush boom towns was a village located about five and a half miles south of where Dahlonega now stands. In June of 1832, William Dean built a cabin here on a hill situated between the valleys of the Chestatee and Etowah Rivers. Shortly thereafter, Nathaniel Nuckolls moved in and erected a small tavern and hotel. The establishment was so popular among the miners that it quickly expanded and the place came to be called Nuckollsville, sometimes spelled Knucklesville. The latter spelling was usually a simple mistake, but many people claimed that it was a reflection of frequent brawls that took place among the rough and tumble miners.

The population of Nuckollsville grew rapidly, especially after the lottery began in October. Within a year, according to *Niles' Register*, the town contained "one hundred family dwellings, eighteen or twenty stores, twelve or fifteen law offices, and four or five taverns." By May 1833 the town's population was estimated at one thousand and there was a total of about 10,000 people in the county.

As the town grew, its people became increasingly dissatisfied with the "Knucklesville" designation. Several alternatives, such as Aureola and Aldoradda, were suggested. Some joked that the town might appropriately be called "Scuffle-town" or "Chuckluck City." The name Auraria, from the Latin *aurum* meaning gold, was selected by a local citizen named John Powell, though the honor has sometimes mistakenly been attributed to Senator John C. Calhoun of South Carolina, who owned a local mine and was a frequent visitor to the region.

With its new name and burgeoning population, Auraria seemed to have a bright future. The lottery was scheduled to end in May. Auraria, the only sizable town in the area, had served as the provisional county seat and was almost certain to be made the official center of county government. However, shortly before the Inferior Court met to designate the county seat, a serious problem arose. It was learned that the 40-acre lot upon which most of Auraria stood had been won in the lottery by a man named John R. Plummer, whose right to participate in the drawing had been brought into question by Nathaniel Nuckolls. Facing the need to establish a stable county government, the court had little choice. On April 24,

1833, it announced the selection of a site for the new courthouse north of Auraria, near the Cane Creek mining activities. The Aurarians protested loudly against losing the county seat. Some went so far as to suggest that Auraria and its immediate vicinity secede from the rest of the county. Nevertheless, that summer, county offices and related businesses relocated to the new county seat. Situated on Lot 950 and referred to at the time simply as "Licklog," the town soon became known as Dahlonega.

Georgia gambles away her "rich inheritance" in the land lottery of 1832-33.

The name Dahlonega, from the Cherokee "dalanigei" meaning gold, was indeed appropriate. So abundant was gold in the area that boys often collected sand that washed into the town streets after a shower and panned out small flakes of gold. With its establishment as the county seat, Dahlonega experienced the rapid development that had previously characterized Auraria. Almost overnight, the site was divided into town lots selling for between two and five hundred dollars each. Within a few months nearly a thousand people were crowded into the settlement, with about five thousand people in the surrounding area.

Mining in the Gold Rush Days

Most of the people were directly involved in the hunt for gold. With pans and shovels, they trooped up and down the valleys and hollows looking for a promising spot. These early miners were engaged in what they called "deposit" mining, or what is generally known as placer mining, looking for gold that has washed down from the hillsides and has collected along mountain streams.

When a location was found that showed "good color," the miners set up camp and began working the area in earnest. A variety of crude implements were used to wash these deposits in an effort to extract the fine gold particles. Among the most common were the sluice box, the white oak splint basket, and the "gum rocker." These were used in a separating process, usually worked by two men. Rich sand and gravel was shoveled into the sluice box where water washed it down into a splint basket. This basket acted as a screen which allowed only small particles, including the gold, to pass through to the gum rocker. As the gravel was washed from one end of the rocker to the other, the heavier minerals were caught by ridges spaced a few inches apart at the bottom of the contraption. These sediments were collected periodically and the gold was panned out by hand.

Another widely used device was the "Long Tom," which combined the actions of the sluice box, splint basket, and gum rocker. It consisted of two rectangular boxes, one set above the other, separated by a cast iron plate. Holes in the plate allowed smaller particles of dirt to pass through to the lower box, where ridges caught the heavier elements while the rest washed off.

For those with a bit more money to throw into the search for placer deposits, the river bottoms were very attractive. In 1833, miners drifted flatboats into the Chestatee and Etowah Rivers and dredged up sand and gravel. It took two men a whole day to fill a boat, but it was usually worth the effort, since many loads contained from forty to fifty dollars' worth of gold. John C. Calhoun once invested in a scheme to change the course of the Chestatee River in an attempt to get at the treasures which lay at its bed, but the project was cancelled when a local "proprietor" refused to allow the proposed channel to cross his land.

Placer mining was popular among the early gold diggers because it required very little capital, but as more people moved into the gold region and towns began to grow, money became available for investment in vein mining. It was evident that the gold nuggets lying along mountain streams had been washed down from veins embedded in the surrounding hills. Now that the lottery was over and land ownership was more settled, miners began prospecting for the rich gold veins.

The equipment required to conduct vein mining was considerably more expensive than that involved in washing placers. Once tunnels had been dug, rafters had to be put in place to avert cave-ins. Pumps were necessary to clear water from the mines, though existing technology made it impractical to attempt much mining below the water table. Most mines ran no deeper than one hundred feet and in the years since the gold rush only a little over twenty mines have been sunk below that depth.

Once a thriving frontier settlement of more than one thousand people, Auraria is today a virtual ghost town.

The search was spurred on by newspaper editorials expounding the merits of vein mining, pointing out that the placer deposits would eventually give out. Following their advice, mining companies began forming in Habersham County, now White County, as early as 1832. Within a year, the state was regularly granting charters to such ventures as the Cherokee Mining Company, the Chestatee Mining Company, and the Lumpkin County Mining and Manufacturing Company.

When vertical shafts were sunk, equipment had to be set up to haul the gold-bearing ore to the surface. Once the ore was brought out of the mines, it was crushed in a stamp mill. These ranged in size and complexity from a single stamp suspended from a bent-over sapling to mills consisting of ten or more stamps driven by a water wheel. After the ore was crushed almost to a powder, it was washed over a plate lined with "quicksilver," or mercury, to draw out the gold. The gold stuck to the mercury plates and was later separated in a retort vessel.

Much of the ore that was extracted from north Georgia mines was extremely pure. A single miner took several hundred pennyweights of gold from a vein in Cherokee County in just a few hours. The Battle Branch mine, located on the Etowah River, just west of Auraria, yielded about $20,000 a ton during its operational peak, even though the yield of such mines was limited by the inefficiency of the old-style mercury plates, which allowed from 50 to 90 percent of the gold to escape. Even today, north Georgia gold remains some of the purest in the world. Deposits that are 70 parts per 100 gold, or 70 percent fine, are normally considered profitable to mine. Georgia gold averages about 95 percent fine as compared to the 88 percent of California gold.

The old Graham Hotel is the only building left in Auraria that dates from the gold rush era—the last relic of America's first gold rush town. Attempts to attain funding for its restoration have proved unsuccessful.

Miners often used gold as a direct medium of exchange. They carried gold dust in goose quills, small pouches, or simply loose in their vest pockets. On Saturday afternoons, the miners took their gold into town and sold it to the storekeepers or traded it for merchandise. Taverns were very popular among the miners, and they spent much of their gold in these establishments. The going price for a drink was the amount of gold dust that would lie on the point of a knife. Some of the gold made its way to Tennessee and Kentucky in exchange for livestock brought to the gold region by drovers from those states. Gold was also sold to individuals and bankers at very low rates. Because of its fineness, Georgia gold was popular in the making of jewelry and was also in demand as an export commodity.

A substantial portion of the gold was also sent to the federal mint in Philadelphia. Between 1830 and 1837, the Philadelphia Mint received $1,798,900 worth from the Georgia mines. But the journey took time and it was usually more convenient to have one's gold coined at a less distant private mint. Some miners sent their gold to Christopher Bechtler in North Carolina, who coined $3,625,840 between 1831 and 1840.

The best-known of the private mints in Georgia was that of Templeton Reid. In 1830 Reid, a gunsmith and clock maker, established a mint in Milledgeville and then relocated to Gainesville; he coined $2.50, $5.00 and $10.00 gold pieces. Gold brought to the Reid mint was meticulously assayed and the coins were usually worth a bit more than their face value. An 1835 article in the *American Journal of Science and Arts* reported that an analysis of a Reid $10.00 gold piece revealed a fineness of 95.579, the remainder being silver. The Reid mint remained in operation, but it is certain that Reid was out of the business by 1832. It is even less certain how many coins were minted during this short time. The best estimate gives a figure of about 1,500 coins with a face value of approximately six thousand dollars. In any case, the coins are today extremely rare. The Smithsonian Institution has one full set and possibly one other is in private hands. There are probably no more than six pieces of each issue in existence. In 1979 a $5.00 gold piece sold for $200,000.

Life in the Gold Region

Bill and Tom Jenkins sluicing for gold on Tanyard Branch about 1912 in Tate's pasture, now part of the North Georgia College campus.

The people who worked both the placer and vein mines were a varied lot. Some roamed the countryside searching for deposits while others sought employment with newly formed mining companies. It was also common practice for those who owned gold lots to lease their property in exchange for one-fourth of the gold mined. In most instances, especially where absentee landowners were involved, "a standing joke among the miners [was] that they had left the rent in the ground, and the owners could get it themselves when they wanted it bad enough." Some miners did not even bother to inform the owners they were digging on their land. This practice, known appropriately as "swindling," was so common along one little stream that it came to be called Swindling Branch.

Mining was very popular as an off-season activity among farmers and planters. After the fall harvest, many farmers headed for the hills to try their luck in the gold fields. Planters bought or leased gold lands and worked slaves in the mines. At one point, John C. Calhoun had twenty slaves working in his mine. Though some planters found it quite profitable to use slaves in mining, the practice was not widespread. Local residents objected to the presence of a large slave population in the vicinity. Mine operators preferred white labor since it was cheap and did not incur the responsibilities of food, clothing and shelter associated with slavery.

The gold rush also attracted a good many foreigners to north Georgia. They were mainly natives of Germany, Sweden, Switzerland, Spain, and the British Isles, and many had previous experience in the mines of Europe and South America. Two brothers from Ireland, who "understood the science of digging to perfection," leased one or two lots around Dahlonega and within a short time were said to be worth $15,000.

Seawell dredge boat on the Chestatee River near Long Branch, where Georgia highways 60 and 400 meet five miles south of Dahlonega. This photo was taken just after the turn of the century.

Conflicts between the miners and the Cherokee, who held steadfast to their claim on the land, were frequent. One such conflict occurred near a place called Scudder's in Forsyth County. About thirty Cherokee and a like number of whites battled with "fists, sticks and stones." No one was killed, but there were several black eyes and broken bones. Several weeks later, another serious dispute broke out at a mine in Lumpkin County. According to a contemporary newspaper account,

> Messrs. Bolton and Lindsey had purchased the lot which gave rise to the difficulty; and that Lindsey went in possession with a company of twelve or fourteen hands and commenced operating for gold. Our informant also states that there was an Indian house on the lot, and that an Indian by the name of Bean claimed the possession of it, and forbid Lindsey's right to it. Lindsey paid no attention to him until . . . on Tuesday last, the war whoop was raised from the house on the lot,

> and about twenty-five or thirty Indians all painted and undressed, rushed out and continued the yell of war, until they got in proper distance of the white men on the branch, and attacked them with sticks, clubs and rocks. The whites defended themselves with their mining tools; the contest was kept up about two hours. A white man, by the name of Goodwin, lately from Kentucky, got his arm broken in two places, and six or seven Indians [were] dangerously wounded. The whites are still in possession.[9]

Another unusual confrontation occurred when "an almost nude Indian woman of powerful frame" attacked a group of miners who were digging along a stream near Dahlonega. She jumped down into their pit and put up such a fight that the miners felt lucky to escape with their lives. In commemoration of the event, this stream came to be called Amazon Branch.

Still other battles took place between gangs of white miners fighting for possession of choice gold property. On one occasion, sixty Carolinians set upon a group of twenty Georgians. When the attackers were finally driven off, one man was mortally injured and others were seriously hurt. In May of 1831, a confrontation developed on a small tributary of the Etowah River. An argument over property rights broke out between a group of miners from Tennessee and a party of Georgians. The dispute quickly turned violent and a number of men were badly injured. According to some reports, at least one man was killed. To this day the place is known as Battle Branch.

The swelling population caused other problems. The first difficulty encountered by a traveler of the day was simply getting to the gold region. There were few roads, and those were in poor condition. At times, the spring rains made the roads impassable. Even so, stage lines were established in an attempt to keep northern Georgia in touch with the rest of the world. These included a twice-weekly service between Pendleton, South Carolina (John C. Calhoun's hometown) and Gainesville, and a line that ran from Athens to "Lumpkin County Courthouse" three times a week.

Though transportation was rough and unpredictable, the hotels and taverns that sprang up across the region, if not luxurious, were at least comfortable. The proprietor of the Troup Hotel in Clarksville flattered himself "that he can boast of as spacious, commodious and pleasant an establishment, as any in the up-country of Georgia," according to the *Southern Banner* of May 25, 1833. One early traveler wrote of the Dahlonega hotels, "the comforts for the inner man were not excelled by any in the state."

Working a mine tunnel in the 1950s.

One of the most popular of the area's hotels was run by Agnes Paschal, or "Grandma" Paschal as she was affectionately called by the folks in and around Auraria. In January of 1833, one of her sons, George W. Paschal, moved to Auraria and set up a law office. He soon purchased the hotel and tavern of Nathaniel Nuckolls and in April the *Western Herald* announced the opening of "Mrs. Paschal and Sons." The paper assured potential customers that they would receive as good accommodations as any in the country. Grandma allowed no liquor in her establishment; nevertheless, the hotel was quite popular. Important civic meetings were often held there, and John C. Calhoun made the Paschal Inn a regular stop on his visits to Auraria.

So many people moved into north Georgia so quickly that there were shortages of almost everything. To satisfy the demand for meat, farmers from as far away as Tennessee and Kentucky brought their livestock to the Georgia gold country. Hundreds of hogs, sheep, cattle, even turkeys were driven down the Old Federal Road to Pickens County all year round. From there they went to Cherokee, Lumpkin and other north Georgia counties, or on to the markets of Athens, Milledgeville, and Augusta.

Doctors were also in short supply, and few were willing to venture into the rugged countryside at all hours of the day and night. Even worse, a good portion seemed more intent on hunting for gold than ministering to the sick and injured. This healing task was often left to people like Mrs. William R. King and Grandma Paschal. For a number of years there was no doctor in Gilmer County and Mrs. King was frequently called upon to perform the functions of both doctor and nurse. Though there were a few doctors in Auraria, the infirm often preferred the attention of Granny

Paschal. This was due largely to her stern rejection of common medical practices of the day such as leeching, bleeding, and blistering. She relied more upon her own philosophy of employing "herbs, ventilated rooms and patient nursing." Grandma's success rate in treating patients was, not surprisingly, better than average.

Despite scarcity and hardship, the Georgia gold region was an exciting place in the 1830s. Wrote one eyewitness of those early days, "Scarcely a stream in the whole country but what was thronged with miners, delving after the precious metal, while the hills and valleys were made to reverberate with the busy rattle of machinery, and the clinking of picks and shovels."

When the miners were not in the countryside digging for gold, they could usually be found in town looking for ways to spend it. According to one account, "Gambling houses, dancing houses, drinking saloons, houses of ill fame, billiard saloons, and tenpin alleys were open day and night." Very often, arguments would erupt into free-for-alls, which "made night hideous with fighting, cursing and swearing" in the gold rush towns. The men were nearly matched in number by women, who were, wrote one resident of Dahlonega, "equally as vile and wicked."[10] It appears that as a group, the most "moral" people were those denied an opportunity to participate in this "devilment"—the slaves.

For those so inclined, and there were not many in those early gold rush days, religion provided a source of escape from life's harsh realities. When Agnes Paschal arrived in Auraria early in 1833 there was not a single church in the town. She set about organizing one, but the small congregation was not able to secure a regular preacher for some time. Itinerant ministers were occasionally sent into the gold region, but they "frequently forgot the sinner, left Satan to his own devices, laid aside their Bibles, and took up the gold pan in pursuit of the yellow mammon, the love of which is reputed to be the root of all evil."[11]

If the gold country could not attract ministers of the gospel, it had no such difficulty when it came to outlaws. Several gangs of lawless men roamed the area, but serious criminal acts were not as frequent as the popular imagination conjures up. In Auraria, there was only one murder recorded at the height of the gold rush. Most criminal incidents were of a less serious nature and involved such things as horse stealing or other forms of petty robbery. The most famous of the gold rush outlaws was a man who never really existed, Guy Rivers, the fictional creation of William Gilmore

Simms.[12] There are those who say that Guy Rivers was based on a real outlaw who terrorized the gold region in the 1830s. A local legend has it that Guy Rivers buried his ill-gotten gold in a cave just west of Auraria overlooking the Etowah River where, as far anyone knows, it still lies undiscovered.

Panning for gold around 1900 near the Briar Patch Mine, four miles south of Dahlonega.

With the image of men like Guy Rivers in the area, it is not surprising that the miners sought ways of safeguarding their gold. In the early days of the Intrusion, some miners hid their gold under rocks, in hollow trees, or buried it. These forms of "banking" had obvious drawbacks, and it quickly became apparent that the miners needed real banks. The Pigeon Roost Mining Company opened the region's first bank in Auraria. The company issued notes for use as payment to their employees, and it is certain that many of these notes circulated in the county before being presented at the Pigeon Roost office for conversion to gold. Soon regular commercial banks from across the state began to establish branches in the gold region. The Bank of Darien set up offices in Clarksville, Auraria, and Dahlonega. Clarksville's own "Farmers, Bank of Chattahoochy" had a branch office in Lumpkin County.

The Dahlonega Gold Mint

Price Memorial Hall, the administration building of North Georgia College. This building was constructed in 1879 on the stone foundations of the old mint after it was destroyed by fire. In 1973, to celebrate the college's 100th anniversary, the steeple was leafed with gold from the surrounding hills.

Even with the establishment of banking, trade in the gold region suffered from the lack of a suitable medium of exchange. Only about one coin per person circulated in the area, and credit was difficult to come by. Merchants insisted on immediate payment in cash or gold, and it was the merchants who determined the gold's worth. What the miners needed was a local mint where they could have their gold assayed and exchanged for gold coins. A proposal for just such a mint was drawn up at a meeting held in the Paschal Hotel in November 1833, and it immediately received the enthusiastic support of Senator Calhoun. Two years later, Congress authorized the establishment of a federal branch mint at Dahlonega, and in 1838 the new mint went into operation. It coined over $100,000 worth of gold in its first year, and by the time it closed in 1861, it had produced almost one and a half million gold coins with a face value of over six million dollars. But for all this, the mint served not a single purpose for which it was intended. It did not produce enough to help overcome the national shortage of coins, as had been hoped. It did not even smooth out the local economy, since gold coins did not freely circulate in the area (they were considered much too valuable). The facility was kept in operation only because of its political popularity.

When the mint went into operation, many saw it as a national affirmation of Georgia's efforts over the preceding decade. To the area's residents, this federal establishment was a seal of approval and a promise of future prosperity. There was but one task left to be performed to make the transition to a new age complete: removal of the Cherokee.

The Trail of Tears

Following Georgia's response to the *Worcester* decision, some Cherokee felt that the battle to save their homes was suicidal and it would be best to conclude as favorable a treaty of removal as possible under the circumstances. In December of 1835 a minority faction of the Cherokee Nation, including Major Ridge, his son John, Elias Boudinot, and Stand Watie, gathered at New Echota to place their signatures on a removal treaty. During preliminaries to the signing, Major Ridge delivered his position in eloquent and moving terms.

> I am one of the native sons of these wild woods. I have hunted the deer and turkey here more than fifty years. I have fought your battles, have defended your truth and honesty, and fair trading. I have always been the friend of honest white men. The Georgians have shown a grasping spirit lately; they have extended their laws, to which we are unaccustomed, which harass our braves and make the children suffer and cry; but I can do them justice in my heart. I know the Indians have an older title than theirs. We obtained the land from the living God above. They got their title from the British. Yet they are strong and we are weak. We are few, they are many. We cannot remain here in safety and comfort. I know we love the graves of our fathers. We can never forget these homes, I know, but an unbending, iron necessity tells us we must leave them. I would willingly die to preserve them, but any forcible effort to keep them will cost us our lands, our lives and the lives of our children. There is but one path to safety, one road to future existence as a Nation.[13]

This Treaty of New Echota called for the Cherokee to relinquish all claims to their eastern lands in exchange for five million dollars and the hollow promise that their new lands in the west would never become part of another state or territory without Cherokee consent.

A majority of the Cherokee, including Chief John Ross, rejected the Treaty of New Echota and denounced those who signed it as nonrepresentative of the official Cherokee position. Nonetheless, the United States government made it clear that if the treaty's terms were not carried out within two years of its signing, federal troops would forcibly evict the Cherokee from their homes.

While the government made preparations to escort the Cherokee west of the Mississippi, efforts to forestall the seemingly inevitable trek continued. Ross and his delegation spent most of 1836-37 in Washington petitioning

John Ross, chief of the Cherokee, held out until the end against removal. His wife, Qatie, died on the Trail of Tears.

Congress to reconsider the Indian Removal Act. Such prominent figures as Ralph Waldo Emerson argued for the Cherokee cause. In an open letter to President Martin Van Buren, Emerson warned that if the United States persisted in its efforts to remove the Indians, "the name of this nation, hitherto the sweet omen of religion and liberty, will stink to the world."[14]

Questions of reason and morality notwithstanding, in April 1838 Van Buren ordered Major General Winfield Scott to north Georgia with five regiments of regulars and four thousand militia and volunteers. In May they began rounding up the Cherokee and herding them into stockades in preparation for removal. A brief account of the scenes that occurred were vividly recorded by James Mooney.

> Families at dinner were startled by the sudden gleam of bayonets in the doorway and rose up to be driven by blows and oaths along the weary miles of trail that led to the stockade. Men were seized in their fields or going along the road, women were taken from their [spinning] wheels and children from their play. In many cases, in turning for one last look as they crossed the ridge, they saw their homes in flames, fired by the lawless rabble that followed on the heels of the soldiers to loot and pillage.[15]

By autumn most of the Cherokee were in custody and the long journey to what would later become Oklahoma began. Of more than thirteen thousand people who set out on the march, nearly five thousand died along the way. Small wonder that this forced migration came to be known as the Trail of Tears. One Georgia volunteer who was later a colonel in the Confederate army wrote of it: "I fought through the Civil War and have seen men shot to pieces and slaughtered by thousands, but the Cherokee removal was the cruelest work I ever knew."[16]

The Gold Plays Out

Tiny specks of gold can still be seen in the bricks of the old courthouse. Completed in 1836, it has served as the Dahlonega Gold Museum since 1967.

The past is filled with ironies, one of the most profound of which is seen in the climax of the North Georgia Gold Rush. For just as the mint began production and the last of the Cherokee were removed, the reason for both began to disappear. The gold rush was over.

By the early 1840s, it was becoming increasingly difficult for the miners to make a living washing the placer deposits. The mining companies, too, felt the pressure of diminishing profits. Although best estimates place the total production of gold in Georgia from 1828 to 1837 at around twenty million dollars, between 1838 and 1849 the gold fields yielded only about one-fourth that amount. Nevertheless, a good portion of the miners clung to their dreams of gold and continued panning the rivers and branches of northern Georgia through the 1840s.

By 1849, word of the great California strikes had reached the gold-hungry miners of Georgia and they began making ready for the long journey westward. Dr. Matthew Stephenson, amateur geologist and assayer at the Dahlonega mint, called a meeting at the town square to discuss the matter. Mounting the courthouse steps to address a crowd of around two hundred miners, Stephenson chastised them for allowing their heads to be turned by fantasies of gold in California. Waving his hand toward Findley Ridge, just south of Dahlonega on Crown Mountain, Stephenson shouted, "Why go to California? In that ridge lies more gold than man ever dreamt of.

There's millions in it." Despite Stephenson's admonitions, the miners left for California in droves, but they took his parting phrase with them. "There's millions in it" became a well-known saying among the California miners and it kept them digging there even if it did not inspire hope on Findley Ridge. This phrase, which Mark Twain later attributed to his "enthusiastic" character, Mulberry Sellers, gave rise to the more widely known cry, "Thar's gold in them thar hills."

The 1850s saw a renewed interest in gold production with the advent of hydraulic mining. This process, first developed in California, involved the construction of canals to transport water to the mines. The water was funneled into a giant hose and the elevation of the canals provided enough force to wash tons of dirt from the mountains. Several hydraulic mining companies were formed, and by the end of the decade, great gashes were being cut in the north Georgia hillsides. However, the coming of the Civil War in 1861 closed the mint and brought mining in the area to a virtual standstill.

A New Beginning

Hydraulic mining in Lumpkin County around 1912.

After the war, mining continued sporadically, and several hydraulic operations were set up. But significant interest in gold mining was not seen again until the turn of the century when five companies set up gold-processing plants around Dahlonega. In June of 1899, the Dahlonega Consolidated Gold Mining Company began construction of a 120-stamp mill on the banks of Yahoola Creek just east of town. The main building measured 300 by 100 feet and was four stories tall, making it the largest gold plant ever built in the United States east of the Mississippi. Despite huge capital investments, the processing plants could not be made to turn a profit and by 1906 they had all been forced to close.

Another period of gold fever struck in the 1930s. Several mines, including the Franklin in Cherokee county, the Thompson in White County, and the Lockhart, Battle Branch, and Pigeon Roost (Barlow) in Lumpkin County, were briefly worked. But these proved to be no more productive than the mines of thirty years earlier. In 1939 Graham Dugas reopened the old Calhoun Mine. He made a good deal of money convincing investors to finance the scheme, but the mine itself produced little wealth.

Since the 1930s, there have been no serious attempts to mine the gold of north Georgia on a large scale, but gold fever is just as strong in the region today as it ever was. Every October during Gold Rush Days, Dahlonega is thronged by modern-day prospectors, reliving the days when gold nuggets could be found in the streets. There is still gold in the Georgia hills. Almost any shovel of dirt turned over in the gold belt will yield a few tiny, glittering flakes. Some geologists have recently predicted that the really big strike has yet to be made. This comes as no news to the old-timers. They've been saying that for years.

Notes

1. Lucian Lamar Knight Clipping Scrapbook, Georgia Department of Archives and History, Atlanta, 142.

2. Larry E. Mitchell, "Benjamin Parks: A Really Golden Heritage," *North Georgia Journal 2* (1) (1985): 23.

3. *Atlanta Constitution*, July 15, 1894.

4. George R. Gilmer, *Sketches of Some of the First Settlers of Upper Georgia, of the Cherokees, and the Author* (Americus: Americus Book Co., 1926), 265.

5. James W. Covington, ed., "Letters from the Georgia Gold Region," *Georgia Historical Quarterly* 39 (1955): 407-8.

6. Howard Zinn, A *People's History of the United States* (New York: Harper and Row, 1980), 134.

7. Gilmer, *First Settlers*, 285.

8. Wilson Lumpkin, *Removal of the Cherokee Indians from Georgia, 1827-1838*, 2 vols. (Savannah: The Savannah Morning News Print, 1907), 1: 128.

9. *Western Herald*, quoted in the *Southern Banner*, May 18, 1833.

10. William P. Price, "Sixty Years in the Life of a Country Village Baptist Church," in Cain, *History of Lumpkin County*, 228.

11. Cain, *History of Lumpkin County*, 47.

12. William Gilmore Simms, *Guy Rivers: A Tale of Georgia*, 2 vols. (New York: Harper and Brothers, 1834).

13. Thurman Wilkins, *Cherokee Tragedy: The Ridge Family and the Decimation of a People* (Norman and London, University of Oklahoma Press, 1986), 286-87.

14. Ralph Waldo Emerson, *Complete Works* (Boston: Houghton Mifflin, 1903–04), 11:95–96.

15. James Mooney, *Myths of the Cherokee*, Bureau of American Ethnology 19th Annual Report, Part 1 (Washington, D.C.: Government Printing Office, 1900), 130.

16. Mooney, *Myths of the Cherokee*, 130.

Discussion Questions

1. Now that you have seen this documentary, do you think it should have a different title? If yes, what would you call it?

2. If you had been living in north Georgia during this period, do you think you are the kind of person who would have caught "gold fever"?

3. Did anything in the video really surprise you or shock you?

4. Do you think the writer of the script had a particular social, historical, philosophical, or political message to convey (an ax to grind)?

5. Would you enjoy pursuing the topics introduced in "Gold Fever" further, or are you satisfied that what was presented quenches your historical thirst about this episode of Georgia's past?

6. Do you think the fate of the Cherokees might have been different without the gold rush?

7. Do you think the kind of people who get caught up in adventures like a gold rush are unique individuals?

8. Do you think the behavior of people who represented state and federal governments as described is pretty much the same as it would be today? Or are government officials better or worse today?

9. Why do you think panning for gold was so contagious for so many people? Is it simply the possibility of "striking it rich"?

10. The video production presented women who indicated experiencing as much "gold fever" as the men. Is this what you would have expected?

11. Was the story of the gold rush improved by the interviews with descendants of people who were a part of this history?

12. If you have been to Dahlonega, do you feel that the city does a good job of introducing visitors to its heritage, especially as it compares with the information presented in the documentary?

13. If you have not been to Dahlonega, would you like to visit the area now that you have learned about it from the documentary?

14. If you could step back in time to this period of Georgia's history and live in the north Georgia mountains, what kind of work would you find most exciting? Fox example, sheriff, tavern owner, hotel operator, barber, newspaper editor, minister, farmer, miner, mint worker, shopkeeper, or some other occupation?

These discussion questions were prepared by Dr. Margaret Holt of the Department of Adult Education, University of Georgia.

Suggestions for Further Reading

The Dahlonega Consolidated Gold Mining Company, which operated betweeen 1899 and 1906, was the largest gold mining plant built east of the Mississppi River.

By far the most detailed and comprehensive treatment of Georgia's gold rush era is H. David Williams, "The North Georgia Gold Rush" (Ph.D. dissertation, Auburn University, 1988). The work extensively integrates the themes of the Great Intrusion, the Lottery of 1832, settlement of the region, and Cherokee removal. The best short overview of antebellum gold mining in Georgia is Fletcher M. Green's "Georgia's Forgotten Industry: Gold Mining," *Georgia Historical Quarterly* 19 (1935): 91-111, 210-28. A less scholarly, but still valuable, overview of north Georgia and gold is the beautifully illustrated little book by Lou Harshaw, *The Gold of Dahlonega* (Asheville, N.C.: Hexagon Co., 1976).

A good portrait of life in North Georgia at the height of the gold rush is E. Merton Coulter's *Auraria: The Story of a Georgia Gold Mining Town* (Athens: University of Georgia Press, 1956). Based largely on the *Western Herald*, Auraria's newspaper in 1833-34, this short book is a window on the era as experienced by residents of America's first gold rush town. In dealing with such people as doctors, lawyers, merchants, barbers, preachers, and schoolmasters as well as miners, *Auraria* comes closer than any other work to giving the reader a feel for what it might have been like to live in that place and time.

Some brief treatment of gold mining may be found in county histories such Lloyd G. Marlin's *History of Cherokee County* (Atlanta: Walter W. Brown

Publishing Co., 1932) and George G. Ward's *Annals of Upper Georgia Centered in Gilmer County* (Nashville: The Parthenon Press, 1965). However, the only such history to devote large sections to gold mining is Andrew W. Cain's *History of Lumpkin County for the First Hundred Years, 1832-1932* (Atlanta: Stein Publishing Co., 1932; reprint ed., Spartanburg, S.C.: Reprint Co., 1984). The book's primary value to the student of Georgia gold mining is twofold. First, it is a marvelous sourcebook. Cain includes verbatim texts of articles from Dahlonega newspapers, some of which might otherwise be lost, and the work contains interviews with people who remember what those who lived through the gold rush said about it. Secondly, Cain provides a look at gold mining in north Georgia during the late nineteenth and early twentieth centuries, with particular emphasis on the turn-of-the-century boom period.

The best of contemporary accounts is *Ninety-Four Years, Agnes Paschal* (Washington: n.p., 1871; reprint ed., Spartanburg, S.C.: Reprint Co., 1974) by her son, George W. Paschal. The book deals primarily with the life of "Granny" Paschal, but the second half gives the reader a personal view of the woman in the setting of her times. It is as much a social history as a biography. Other contemporary accounts include George R. Gilmer's *Sketches of Some of the first Settlers of Upper Georgia* (Americus, Ga.: Americus Book Co., 1926) and a collection of Wilson Lumpkin's writings called *Removal of the Cherokee Indians from Georgia, 1827-1838* (2 vols., Savannah: The Savannah Morning News Print, 1907). These works give the perspective of the two gold-rush-era governors of Georgia. As might be expected, neither Gilmer nor Lumpkin speaks highly of the Cherokee, though Lumpkin's prejudicial disdain is slightly less obvious.

A standard work dealing with the Cherokee and their civilization in the early nineteenth century is James Mooney's *Myths of the Cherokee* (Bureau of American Ethnology 19th Annual Report, Part 1; Washington, D.C.: Government Printing Office, 1900). This treatment is as valuable for Mooney's insights as for the numerous personal accounts of the battle with Georgia and the Trail of Tears. More recent works of value on the Cherokee during the gold rush include Gary E. Moulton's *John Ross: Cherokee Chief* (Athens: University of Georgia Press, 1978) and Thurman Wilkins' *Cherokee Tragedy: The Ridge Family and the Decimation of a People* (Norman: University of Oklahoma Press, 1986).

Two excellent books on the Dahlonega Mint deserve attention. C. M. Birdsall's *The United States Branch Mint at Dahlonega, Georgia: Its History and Coinage* (Easley, S.C.: Southern Historical Press, 1984) is primarily a statistical examination of the mint's production, while *The Neighborhood Mint: Dahlonega in the Age of Jackson* (Macon, Ga.: Mercer University Press, 1986), by Sylvia Gaily Head and Elizabeth W. Etheridge, is a portrait of the Mint in the context of its time. Both works are extensively annotated.

The North Georgia Gold Rush was by no means an isolated phenomenon in the South. For those interested in gold mining in other southern states, the following articles are recommended:

Otis Young, "The Southern Gold Rush, 1828-1836," *Journal of Southern History* 48 (August 1982): 373-92; Robert A. Russell, "Gold Mining in Alabama Before 1860," *Alabama Review* 10 (1957): 5-14; and two by Fletcher M. Green, "Gold Mining in Antebellum Virginia," *Virginia Magazine of History and Biography* 45 (1937): 227-35, 357-66, and "Gold Mining: A Forgotten Industry of Antebellum North Carolina," *North Carolina Historical Review* 14 (1937): 1-19, 135-55.

Finally, there may be some who would like to try their hand at panning for gold. W. Larry Otwell's book *Panning Georgia's Gold* (Cleveland, Ga.: Rainbow Sequoia Pub., 1985) is a complete guide to all aspects of amateur gold hunting, including proper equipment, how to find a promising spot, and etiquette involved in panning on private as well as public property. There may not be "millions in it," but weekend prospectors will definitely find panning for gold a fascinating, and maybe even a rewarding hobby.

Photo Credits

Page 1 Georgia Department of Archives and History.

Page 6 From "Gold Mining in Georgia," *Harper's Magazine* (1879).

Page 8 From "Gold Mining in Georgia," *Harper's Magazine* (1879).

Page 11 Original painting owned by Rev. Robert L. Whittenburg of Washington, D.C.

Page 18 Georgia Department of Archives and History.

Page 20 Photograph by David Williams.

Page 21 Photograph by David Williams.

Page 23 Georgia Department of Archives and History.

Page 24 Georgia Department of Archives and History.

Page 26 Georgia Department of Archives and History.

Page 28 Georgia Department of Archives and History.

Page 29 Photograph by David Williams.

Page 32 From Thurman Wilkins, *Cherokee Tragedy*. Original located in Philbrook Art Center, Tulsa, Oklahoma.

Page 33 Photograph by David Williams.

Page 35 Georgia Department of Archives and History.

Page 41 Georgia Department of Archives and History.

About the Authors

Raymond C. Rensi is a historian and head of the Social Science Department at North Georgia College in Dahlonega, where he has taught since 1966. He holds a master's degree in history from East Tennessee State University and a Ph.D. in history from the University of Georgia. He has been active in research and writing about the history of north Georgia for many years.

H. David Williams, assistant professor of history at Valdosta State College, received a Ph.D. in history from Auburn University. His doctoral dissertation was "The North Georgia Gold Rush." He has lectured and written articles about the Georgia gold rush and about the experiences of private soldiers during the Civil War. He attended North Georgia College in Dahlonega and earned master's degrees at Washington College, Maryland (history), at Central Michigan University (public administration), and at North Georgia College (education).